THINKING GOD'S THOUGHTS AFTER HIM

GREAT SCIENTISTS WHO
HONORED THE CREATOR

CHRISTINE DAO

INSTITUTE
for CREATION
RESEARCH

Dallas, Texas
www.icr.org

Christine Dao serves as Assistant Editor at the Institute for Creation Research. She is a frequent contributor to ICR's monthly magazine, *Acts & Facts*, and to ICR's Daily Science Updates online, as well as oversees production of the quarterly devotional booklet *Days of Praise*. Ms. Dao is a graduate in journalism from Southern Methodist University in Dallas, Texas.

THINKING GOD'S THOUGHTS AFTER HIM
GREAT SCIENTISTS WHO HONORED THE CREATOR
BY CHRISTINE DAO

First printing: January 2009

Copyright © 2009 by the Institute for Creation Research. All rights reserved. No portion of this book may be used in any form without written permission of the publisher, with the exception of brief excerpts in articles and reviews. For more information, write to Institute for Creation Research, P. O. Box 59029, Dallas, TX 75229.

ISBN-13: 978-0-932766-92-2
ISBN-10:   0-932766-92-7

Reviewers: Randy J. Guliuzza, P.E., M.D., and Brian Thomas, M.S.

Please visit our website for other books and resources: www.icr.org.

Printed in the United States of America.

Contents

Introduction 4

Galileo Galilei 5

Johann Kepler 6

Robert Boyle 7

Isaac Newton 8

Charles Bell 10

William Kirby 11

Michael Faraday 12

James Clerk Maxwell 13

Gregor Mendel 14

Louis Pasteur 15

George Washington Carver 16

Henry M. Morris 17

For Further Study 19

Thinking About Galileo Galilei 21

Thinking About Johann Kepler 22

Thinking About Robert Boyle 23

Thinking About Isaac Newton 24

Thinking About Charles Bell 25

Thinking About William Kirby 26

Thinking About Michael Faraday 27

Thinking About James Clerk Maxwell 28

Thinking About Gregor Mendel 29

Thinking About Louis Pasteur 30

Thinking About George Washington Carver .. 31

Thinking About Henry M. Morris 32

Thinking God's Thoughts After Him
Great Scientists Who Honored the Creator

Introduction

One of the most serious fallacies of modern thought is the widespread notion that biblical Christianity is in conflict with true science and, therefore, that genuine scientists cannot believe the Bible. The scientific method is built on empirical testing of hypotheses, and since creation and other biblical doctrines cannot be tested in the laboratory, they are considered nonscientific, to be taken strictly on faith. Furthermore, it is commonly believed that the Bible contains many scientific errors. At the very most, it is contended, a scientist may be able to accept the spiritual teachings of the Bible if he wishes, but never its scientific and historical teachings.

Such a charge is tragically wrong, however, and has done untold damage. Thousands of scientists of the past and present have been and are Bible-believing Christians. As a matter of fact, the most discerning historians and philosophers of science have recognized that the very existence of modern science had its origins in a culture at least nominally committed to a biblical basis, and at a time in history marked by a great return to biblical faith.

As a matter of fact, authorization for the development of science and technology was specifically commissioned in God's primeval mandate to Adam and Eve (Genesis 1:26-28), and many early scientists, especially in England and America, viewed it in just this way. The study of the world and its processes is really, as Johann Kepler and other great scientists have maintained, "thinking God's thoughts after Him," and should be approached reverently and humbly.

In this book, we present a number of brief biographical testimonies of important scientists who professed to be Bible-believing Christians. Many of these names will be names familiar to every science student, but he or she may not know that these men also were Christians (this fact is commonly ignored or slighted in present-day scientific literature). This will by no means be an exhaustive list, but it should at least put to rest the common misconception that no first-class scientist can be a Bible-believing Christian.

Some of these scientists lived before the rise of modern Darwinism, but they were certainly well aware of evolutionary philosophy (which has been around since antiquity) and of scientific skepticism in general (deism, humanism, atheism, pantheism, and other antibiblical philosophies were very real threats to Christian theism long before the modern era). Nevertheless, they were all convinced of the authority of Scripture and the truth of the Christian worldview.

Like people in other professions (even preachers), scientists have held a variety of specific religious beliefs. The inclusion of a particular scientist in this collection will not indicate that we would or would not endorse his personal behavior or particular doctrinal or denominational beliefs. Our only criteria will be that, in addition to being a highly qualified scientist, he believed in the inspiration and authority of the Bible, accepted Jesus Christ as the Son of God, and believed in the one true God of the Bible as the Creator of all things. They will also be seen to represent many different fields of science. In other words, there have been leading scientists in every field of science who have studied both the Bible and their own scientific disciplines in depth, and who are firmly convinced the two are fully compatible.

Adapted from Morris, Henry M. 1988. *Men of Science, Men of God*. Green Forest, AR: Master Books, 1-3.

Man of Science, Man of God:
Galileo Galilei

Who: Galileo Galilei
What: Father of Modern Science
When: February 15, 1564 – January 8, 1642
Where: Pisa, Tuscany

Galileo, whom Albert Einstein called the father of modern science,[1] was born Galileo Bonaiuti de' Galilei in Tuscany on February 15, 1564. Although he had considered entering the priesthood, his father sent him to study medicine at the University of Pisa. He switched his focus to mathematics, becoming mathematic chair in Pisa in 1589. In 1592, he moved to the University of Padua, where he taught mechanics, geometry, and astronomy for the next 18 years.

Galileo made significant scientific contributions based on mathematics and experimentation. Among his many achievements, he developed new and more powerful telescopes, and in 1610 he published his telescopic observations in *Sidereus Nuncius (Starry Messenger)*. He also discovered four of Jupiter's largest moons, which were later named the Galilean satellites in his honor. He corresponded, and argued, with fellow astronomer Johann Kepler, as well as tutored the young Robert Boyle. In 1619, he disputed with a Jesuit math professor about the nature of comets, and in 1623 published *The Assayer*, presenting views on how science should be practiced.

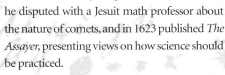
Galileo's telescope

Though a devout Roman Catholic, Galileo was frowned upon by that church for his support of the Copernican theory that the earth revolved around the sun, which was contrary to the accepted Ptolemaic and Aristotelian theories. Galileo didn't think that heliocentrism conflicted with Scripture and published *Dialogue Concerning the Two Chief World Systems* with the intention of presenting both arguments.

The book wasn't well-received. Galileo was suspected of heresy and testified before the Roman Inquisition in defense of his ideas. He was prohibited from publishing, was sentenced first to prison and then to house arrest, and was forced to "abandon the false opinion that the Sun was the centre of the universe and immovable, and that the Earth was not the centre of the same and that it moved."[2] Despite this, in 1638 he published *Discourses and Mathematical Demonstrations Relating to Two New Sciences* beyond the Inquisition's jurisdiction in Holland. He passed away in 1642 and was interred at the Basilica di Santa Croce in Florence, where Michelangelo, Niccolo Machiavelli, and Rossini are also buried.

Galileo's trial is sometimes treated as a case of science vs. religion and used as an example of the "repressive" nature of religious belief. Actually, the situation was more one of experimental science vs. Greek philosophy. Many of Galileo's opponents were disgruntled academics and professors. The opposition to his work shows the need for scientists to be free to explore where the evidence leads, rather than having to conform to the general consensus—a reflection of today's fight for academic freedom against the stranglehold of evolutionary theory.

Another common misconception among secular scientists is that Galileo advocated that science and Scripture should be treated separately, stemming from a letter in which he wrote,

> I believe that the intention of Holy Writ was to persuade men of the truths necessary to salvation....But I do not think it necessary to believe that the same God who gave us our senses...would have us put aside the use of these, to teach us instead such things as with their help we could find out for ourselves, particularly in the case of these sciences, of which there is not the smallest mention in Scripture.[3]

In fact, Galileo's words reflect Proverbs 25:2: "It is the glory of God to conceal a thing: but the honour of kings [or scientists] is to search out a matter." Galileo was a true creation scientist in posing that God created our world and our gifts of reasoning, sense, and understanding to spur us to explore and make discoveries within that creation.

References
1. "Propositions arrived at by purely logical means are completely empty as regards reality. Because Galileo realized this, and particularly because he drummed it into the scientific world, he is the father of modern physics—indeed, of modern science altogether." Einstein, A. 1954. *Ideas and Opinions*. New York: Crown Publishers, 271.
2. Galilei, G. et al. 1870. *The Private Life of Galileo*. Boston: Nichols and Noyes, 299.
3. Ibid, 85.

Background image: Galileo Galilei's discovery of the moons of Jupiter. This is a page, in Italian, from Galileo's published discovery of the moons, which appeared in Sidereus Nuncius *in March 1610.*

Man of Science, Man of God: Johann Kepler

Who: Johann (or Johannes) Kepler
What: Father of Physical Astronomy
When: December 27, 1571 – November 15, 1630
Where: Born in Weil der Stadt, Württemburg, Holy Roman Empire, of German nationality

Johann Kepler developed a love for astronomy at an early age. He observed the Great Comet of 1577 when he was six and the 1580 Lunar Eclipse, events that no doubt fueled his curiosity and enthusiasm for science. Although he originally wanted to be a minister and studied theology at the University of Tübingen, Kepler accepted a position in 1594 as a mathematics and astronomy teacher at a Protestant school in Graz, Austria. He later became an assistant to Tycho Brahe, the court mathematician to Emperor Rudolf II. Upon Tycho's death, Kepler inherited his position, as well as his extensive archive of planetary observations.

Kepler is best known for discovering the three mathematical laws of planetary motion ("Kepler's Laws") that established the discipline of celestial mechanics. He also discovered the elliptical patterns in which the planets travel around the sun. At a time when the sun and other celestial bodies were still widely believed to circle the earth (geocentrism), Kepler defended Nicolaus Copernicus' theory that planets orbit the sun (heliocentrism) and sought to reconcile it with Scripture.[1] He revolutionized scientific thought by applying physics (then considered a branch of natural philosophy) to astronomy (seen as a branch of mathematics).

An "unorthodox" Lutheran, Kepler had a deep love for Christ and the inspiration and authority of Scripture. He is frequently quoted as saying, "O God, I am thinking Thy thoughts after Thee." Strong theological convictions prompted him to find a connection between the physical and the spiritual, and his scientific discoveries led him to believe he had uncovered God's geometrical plan for the universe. In Kepler's view, the universe itself was an image of God, with the sun corresponding to the Father, the stellar sphere to the Son, and the intervening space to the Holy Spirit.

Life, however, held many trials for Kepler. His Protestant beliefs won him little favor with the Catholic church, and the Lutheran church shunned him for his sympathies with Calvinist beliefs. He was forced to relocate more than once to avoid persecution, as well as to escape political dangers from ongoing wars. He suffered the deaths of his first wife and several young children. In addition, fellow scientists did not immediately accept his scientific discoveries. Galileo Galilei and the French mathematician and scientist René Descartes ignored his 1609 work *Astronomia nova (A New Astronomy)*. Even his mentor Michael Maestlin objected to his introduction of physics into astronomy.

Yet Kepler stayed true to his faith, as evident in his written works, and his scientific discoveries would eventually win him acclaim, legitimize the discoveries of his contemporary Galileo, and serve as a major influence on the scientists who came after him. His famous work *Harmonies of the World* (in Latin, *Harmonices Mundi*) begins:

> I commence a sacred discourse, a most true hymn to God the Founder, and I judge it to be piety, not to sacrifice many hecatombs of bulls to Him and to burn incense of innumerable perfumes and cassia, but first to learn myself, and afterwards to teach others too, how great He is in wisdom, how great in power, and of what sort in goodness.[2]

At the end, Kepler concludes:

> Purposely I break off the dream and the very vast speculation, merely crying out with the royal Psalmist: Great is our Lord and great His virtue and of His wisdom there is no number: praise Him, ye heavens, praise Him, ye sun, moon, and planets, use every sense for perceiving, every tongue for declaring your Creator…to Him be praise, honour, and glory, world without end. Amen.[3]

References
1. An extensive chapter in Kepler's *Mysterium Cosmographicum (The Cosmographic Mystery*, the first published defense of the Copernican system) is devoted to reconciling heliocentrism with biblical passages that seem to support geocentrism.
2. Kepler, J. 1619. "Proem." *Harmonies of the World*.
3. "Epilogue Concerning the Sun, By Way of Conjecture," ibid.

Man of Science, Man of God: Robert Boyle

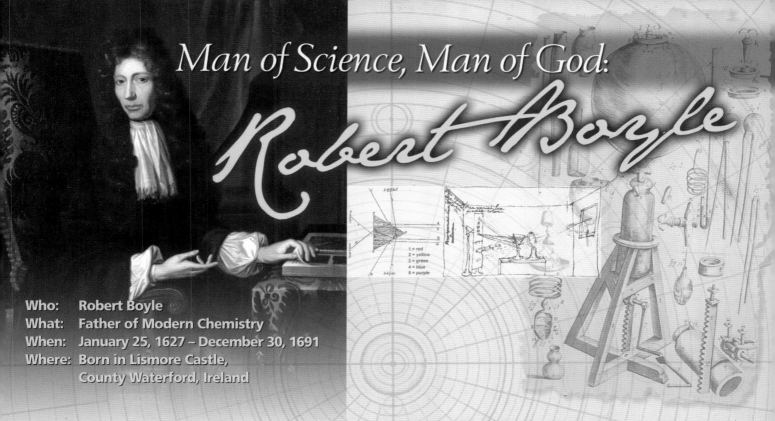

Who: Robert Boyle
What: Father of Modern Chemistry
When: January 25, 1627 – December 30, 1691
Where: Born in Lismore Castle, County Waterford, Ireland

Irish natural philosopher Robert Boyle was a major contributor in the fields of physics and chemistry. One of the first to transform the study of science into an experimental discipline, he also championed the concept that all discoveries should be published, not withheld for personal profit and power—a common practice at the time. A devoted student of the Bible, he also produced multiple books and essays on religion.

The fourteenth child of Richard Boyle, 1st Earl of Cork, young Robert learned to speak Latin, Greek, and French and entered Eton College before he was nine. He later journeyed abroad with a French tutor, including a visit to Florence, Italy, in 1641 to study with the elderly Galileo Galilei. In 1645, Boyle was put in charge of several family estates, marking the beginning of his scientific research. He earned a prominent place in the "Invisible College," a group of scientific minds that were instrumental in forming the Royal Society in 1663.

After moving to Oxford, Boyle and his research assistant Robert Hooke expounded on the design and construction of Otto von Guericke's air pump to create the "machina Boyliana." In 1660, he published his *New Experiments Physico-Mechanical, Touching the Spring of the Air, and its Effects Made, for the most part, in a New Pneumatical Engine*. His response to critics of this work included the first mention of the law that the volume of a gas varies inversely to the pressure of the gas, what many physicists call today "Boyle's Law."[1]

Though he also made discoveries regarding how air is used in sound transmission and the expansive force of freezing water, Boyle's favorite scientific study by far was chemistry, which he believed should no longer be a subordinate study of alchemy or medicine. In 1661, he criticized traditional alchemists and laid the foundation for the atomic theory of matter in *The Sceptical Chymist*, the cornerstone work for modern chemistry.

In addition to his scientific research, Boyle diligently studied the Bible. Along with the Greek he acquired in childhood, he learned Hebrew, Cyriac, and Chaldee so that he could read the text firsthand. His faith drove his experimental studies, as evidenced in his published works, and he believed that science and Scripture exist in harmony. Conflicts between science and the Bible, Boyle explained, were either due to a mistake in science or an incorrect interpretation of Scripture.

> Even when some revelations are thought not only to transcend reason, but to clash with it, it is to be considered whether such doctrines are really repugnant to any absolute catholic rule of reason, or only to something which depends upon the measure of acquired information we enjoy.[2]

His 1681 work *A Discourse of Things Above Reason* stressed the limitations of reason, which Boyle maintained should not be allowed to judge what God's revelation could or could not do. He believed the attributes of God can be seen by studying nature scientifically and that His wisdom is observed in creation.

> When with bold telescopes I survey the old and newly discovered stars and planets when with excellent microscopes I discern the unimitable subtility of nature's curious workmanship; and when, in a word, by the help of anatomical knives, and the light of chymical furnaces, I study the book of nature I find myself oftentimes reduced to exclaim with the Psalmist, How manifold are Thy works, O Lord! in wisdom hast Thou made them all![3]

During his directorship of the East India Company, Boyle promoted Christianity in the East by financially supporting missionaries and translations of the Bible. Upon his death, he endowed a series of lectures in his will designed to defend Christianity. The "Boyle Lectures" are held annually to this day in London, a legacy of this remarkable man of God.

References
1. The original hypothesis was that of Henry Power in 1661 (though Boyle mistakenly attributed it to Richard Townley in his writings).
2. Boyle, Robert. 1690. *Reflections on a Theological Distinction*.
3. Boyle, Robert. 1660. *Seraphic Love*.

Man of Science, Man of God: Isaac Newton

Sir Isaac Newton, perhaps the most influential scientist of all time, came from very humble beginnings. The Julian calendar places his birthday on Christmas 1642, before which his father, John Newton, died at the age of 36. He was born premature and possibly had Asperger syndrome, a form of autism, which could explain his later ability to intensely focus on specific subject matters.

His mother remarried and sent him at age three to live with his maternal grandmother. At 12 he was sent to The King's School, an educational institution for boys in Grantham, Lincolnshire. Biographer N.W. Chittenden recounts that the young Newton was not a good student at first. However, after losing in a fight against the student ranked just above him, he applied himself to his studies until not only did he outrank his offender, but everyone else in his class.

When he was 15, his mother was widowed again and for financial purposes removed him from school to manage a farm. He disliked the work and often neglected his duties, taking advantage of market trips into Grantham to read and study. His mother was persuaded to send him back to school to complete his education.

In 1661 at the age of 18, he entered Trinity College at the University of Cambridge. Newton took an interest in mathematics, overlooking the prominent study of the Greek philosopher Euclid and instead focusing on the relatively modern works of minds such as René Descartes, Galileo Galilei, John Wallis, and Johann Kepler.

In 1665, the young scientist invented the generalized binomial theorem and began developing the mathematical theory that would later become calculus. He received his Bachelor of Arts degree later that year, shortly after which the university was closed as a precaution against the Great Plague. Newton returned to his home in Woolsthorpe to continue his work in calculus, optics, and the law of gravitation, as well as dabbling in some alchemy in the spirit of Robert Boyle's *The Sceptical Chymist*.

He returned to Cambridge in 1668 and earned Master of Arts recognition and the Lucasian Professor of Mathematics position a year

Who: Isaac Newton
What: Father of Universal Gravitation
When: January 4, 1643 – March 31, 1727
Where: Woolsthorpe, a hamlet of Lincolnshire, England

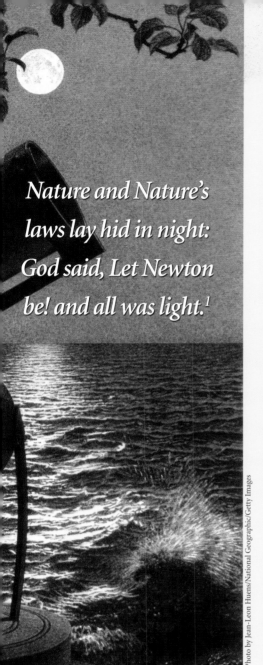

Nature and Nature's laws lay hid in night: God said, Let Newton be! and all was light.[1]

later. The Royal Society took interest in his optics works, particularly his investigations into the refraction of light, as well as the reflecting telescope he invented (today known as a Newtonian telescope). Though his work received initial opposition, it paved the way for Newton's membership into the Royal Society in 1671, sparking the rapid rise of his reputation.

Newton hesitated to publicize his mathematical studies for fear of more opposition. But in 1687, he published the first edition of his *Philosophiæ Naturalis Principia Mathematica* (later translated in 1825 as *The Mathematical Principles of Natural Philosophy*), considered today to be the single greatest work in the history of science. In it, he described universal gravitation and the three laws of motion, derived from Kepler's Laws.

Though he was and still is renowned for his scientific pursuits, Newton was a serious student of the Bible and published several theological works. Even in his famed *Principia*, Newton exhibited his dedication to God.

> This most beautiful system of the sun, planets, and comets, could only proceed from the counsel and dominion of an intelligent and powerful Being....This Being governs all things, not as the soul of the world, but as Lord over all; and on account of his dominion he is wont to be called Lord God παντοκρατωρ, or *Universal Ruler*....[2]

Scientific inquiry, which then existed as Natural Philosophy, could not exist apart from "the Maker," according to Newton. In fact, science was the perfect realm in which to discuss God.

> Since every particle of space is *always*, and every indivisible moment of duration is *every where*, certainly the Maker and Lord of all things cannot be *never* and *no where*....God is the same God, always and every where. He is omnipresent not *virtually* only, but also *substantially*; for virtue cannot subsist without substance....It is allowed by all that the Supreme God exists necessarily; and by the same necessity he exists *always* and *every where*....And thus much concerning God; to discourse of whom from the appearance of things, does certainly belong to Natural Philosophy.[3]

Though he lived before Darwin, Newton was not unacquainted with the atheistic evolutionary theory on origins. He was convinced against it and wrote:

> Blind metaphysical necessity, which is certainly the same always and every where, could produce no variety of things. All that diversity of natural things which we find suited to different times and places could arise from nothing but the ideas and will of a Being, necessarily existing.[4]

In the winter of 1692, Newton suffered the loss of a scientific manuscript 20 years in the making, which triggered a nervous breakdown that lasted almost two years. When he emerged from it, his scholastic work attracted royal attention, and he was appointed as warden and later master of the Royal Mint. Although the appointments were sinecures, he took his work seriously, eventually retiring from his professorship at Cambridge in order to focus on the Mint. He made significant contributions to currency reform and the convictions of counterfeiters and clippers (who clipped the edges of coins, devaluing the currency), crimes considered high treason. His work at the Royal Mint, rather than his scientific achievements, earned him knighthood from Queen Anne in 1705.

Other honors included being elected a member of the French Académie des Sciences in 1699, and becoming President of the Royal Society in 1703. He never married, and though he died without a will, he had already given much of his estate to his nieces and nephews. He also endowed a professorship at the University of Edinburgh in Scotland, and biographers noted that he gave liberally to the poor throughout his life.

Newton died in March 1727 and was interred at Westminster Abbey in London. Although in the popular imagination he is most closely associated with an apple and the law of gravity, Newton himself is quoted as saying, "Gravity explains the motions of the planets, but it cannot explain who set the planets in motion. God governs all things and knows all that is or can be done."[5]

References

1. Epitaph for Newton's grave, composed by English poet Alexander Pope.
2. Newton, I. General Scholium. Translated by Motte, A. 1825. *Newton's Principia: The Mathematical Principles of Natural Philosophy*. New York: Daniel Adee, 501. The Greek word *pantokrator* is most often translated as "Almighty" in the King James Version.
3. Ibid, 505-506.
4. Ibid, 506.
5. Tiner, J. H. 1975. *Isaac Newton: Inventor, Scientist and Teacher*. Milford, MI: Mott Media.

Man of Science, Man of God:
CHARLES BELL

Who: Charles Bell
What: Premier Anatomist and Surgeon
When: November 12, 1774 – April 28, 1842
Where: Born in Edinburgh, Scotland

Sir Charles Bell—anatomist, surgeon, physiologist, natural theologian—was the youngest of four sons. While still a student at the University of Edinburgh, he taught anatomy and published *A System of Dissection Explaining the Anatomy of the Human Body*, which showcased his extraordinarily accurate illustrations. He later published volumes of *Anatomy of the Human Body* with his brother, well-known surgeon John Bell.

Charles' surgical success at the Edinburgh Royal Infirmary and John's popular anatomy classes—and outspoken criticism of the suffering inflicted by incompetent surgeons—led to jealous opposition from local physicians. Eventually the two brothers were barred from practice and teaching in Scotland, and in 1804 they moved to London and opened a private surgery and school of anatomy.

In 1811, Charles published *Idea of a New Anatomy of the Brain*, now considered the "Magna Carta of neurology." After John's death in 1820, Charles continued to teach and conduct research. He was also a surgeon at the Middlesex Hospital, helping to found its medical school in 1828.

Bell had a special interest in the nervous system. For a time, he assumed that all nerves were sensory. His work caught the eye of Francois Magendie, who had demonstrated that the ventral roots of spinal nerves are motor and the dorsal roots are sensory. After a conflict of priority arose, the two scientists reached an agreement and named the rule of spinal nerve function "Bell-Magendie's Law."[1]

Bell's scientific endeavors convinced him of the existence and necessity of the Creator. His 1837 publication *The Hand; Its Mechanism and Vital Endowments, as Evincing Design* affirmed his convictions:

> If we select any object from the whole extent of animated nature, and contemplate it fully and in all its bearings, we shall certainly come to this conclusion: that there is Design in the mechanical construction, Benevolence in the endowments of the living properties, and that Good on the whole is the result.[2]

From studying the human body, Bell realized how dependent people are on involuntary physical processes. He saw close-minded reliance on reason as not only ignorant, but "worse than ingratitude."

> Now, when a man sees that his vital operations could not be directed by reason—that they are constant, and far too important to be exposed to all the changes incident to his mind, and that they are given up to the direction of other sources of motion than the will, he acquires a full sense of his dependence....
>
> When man thus perceives, that in respect to all these vital operations he is more helpless than the infant, and that his boasted reason can neither give them order nor protection, is not his insensibility to the Giver of these secret endowments worse than ingratitude?[3]

Discoveries Named for Charles Bell

Bell's Nerve—the posterior thoracic nerve
Bell's Palsy/Paralysis—a paralysis of facial nerves resulting in inability to control facial muscles on the affected side
Bell's Phenomenon—a medical sign in patients with peripheral facial paralysis that is characterized by the failure of the eyelid on the paralyzed side to close
Bell's Spasm—involuntary twitching of the facial muscles

He was familiar with uniformitarianism, which influenced the development of Darwinism. Bell thought science should be allowed to follow the evidence—even if it leads to a supernatural origin.

> We cannot resist these proofs of a beginning, or of a First Cause. When we are bold enough to extend our inquiries into those great revolutions that have taken place, whether in the condition of the earth, or in the structure of the animals which have inhabited it, our notions of the "uniformity" of the course of nature must suffer some modification.[4]

Bell was made a fellow of the Royal Society in 1826. He later received its first medal, and in 1831 was knighted by King William IV. At age 62, he returned to his homeland as a Professor of Surgery at the University of Edinburgh. His death six years later marked the end of a groundbreaking career in medical science and a humble witness to the intelligent and benevolent handiwork of the Creator.

References
1. The Bell-Magendie Law states that the anterior branch of spinal nerve roots contains only motor fibers and the posterior roots contain only sensory fibers.
2. Bell, Sir Charles. 1852. *The Fourth Bridgewater Treatise on the Power, Wisdom, and Goodness of God as Manifested in the Creation: The Hand; Its Mechanism and Vital Endowments as Evincing Design*, 5th ed. London: John Murray, 1.
3. Ibid, 13-14.
4. Ibid, 265.

Man of Science, Man of God: William Kirby

Who: William Kirby
What: Father of Entomology
When: September 19, 1759 – July 4, 1850
Where: Born in Witnesham, Suffolk, England

William Kirby must have been an unusual fellow, since he derived such joy from studying insects. After graduating from Cambridge University in 1781, he took holy orders in 1782. His interest in natural history was sparked in 1791 when he met English botanist Sir James Edward Smith, with whom he corresponded to seek advice about founding a natural history museum at Ipswich School in Suffolk.

Throughout Kirby's life, he compiled an extensive insect collection. His first major work—*Monographia Apum Angliae*, about the bees of England—caught the attention of leading entomologists in Britain and abroad. He received a Master's degree with the intention of applying for a professorship in botany at Cambridge, but was denied due to his political views.[1]

Between 1815 and 1826, he and fellow British entomologist William Spence co-authored the four-volume *An Introduction to Entomology: or Elements of the Natural History of Insects*. Considered the foundational work in the field of entomology, Kirby introduced it in this way:

> Having given you this full account of the *external* parts of insects, and their most remarkable variations; I must next direct your attention to such discoveries as have been made with regard to their *Internal Anatomy and Physiology*: a subject still more fertile, if possible, than the former in wonderful manifestations of the POWER, WISDOM AND GOODNESS OF THE CREATOR.[2]

> …[W]hen we ascribe a certain degree of intellect to these animals, we do not place them upon a par with man; since all the most wonderful parts of their economy, and those manipulations that exceed all our powers, we admit not to be the contrivance of the animals themselves, but the necessary results of faculties implanted in their constitution at the first creation by their MAKER.[3]

Mankind and the animal kingdom were two distinct creations that shared no ancestors and were defined by wisdom.

> There is this difference between intellect in man, and the rest of the animal creation. Their intellect teaches them to follow the lead of their senses, and make such use of the external world as their appetites or instincts incline them to,—and *this is their wisdom;* while the intellect of man, being associated with an immortal principle, and being in connexion with a world above that which his sense reveal to him, can, by aid derived from heaven, control those senses, and bring under his instinctive appetites, so as to render them obedient to the το ηγεμονιχον, or governing power of his nature: AND THIS IS HIS WISDOM.[4]

In 1835, Kirby authored the seventh *Bridgewater Treatise*, titled *The History, Habits and Instincts of Animals*. The first chapter, "Creation of Animals," argues that the very existence of animals testify to the Creator.

> The infinite diversity of their forms and organs; the nice adaptation of these to their several functions; the beauty and elegance of a large number of them; the singularity of others; the variety of their motions; their geographical distribution; but, above all, their pre-eminent utility to mankind in every state and stage of life, render them objects of the deepest interest…so that arguments in proof of these primary attributes of the Godhead, drawn from the habits, instincts, and other adjuncts of the animal creation, are likely to meet with more universal attention.[5]

With Spence, Kirby helped to found the Entomological Society of London in 1833, to which he was appointed Honorary President for life. His Ipswich natural history museum opened in 1847, and he served as its president until his death in 1850. To this remarkable man, even the insects declared the glory of God.

References

1. Kirby was a Tory, a party that supported the authority of the British monarchy.
2. Kirby, W. 1826. *An Introduction to Entomology: Vol. IV.* London: Longman, Hurst, Rees, Orme, and Brown, 1.
3. Ibid, 32.
4. Ibid, 33.
5. Kirby, W. 1835. *The Seventh Bridgewater Treatise on the Power, Wisdom, and Goodness of God as Manifested in the Creation: The History, Habits and Instinct of Animals*, Vol. 1. London: William Pickering, 1-2.

Man of Science, Man of God: MICHAEL FARADAY

Who: Michael Faraday
What: Father of Electromagnetism
When: September 22, 1791 – August 25, 1867
Where: South London, England

Michael Faraday was arguably the best experimentalist in the history of science. Apprenticed at age 14 to a local bookbinder and seller, he educated himself and developed an interest in science. He finished his apprenticeship in 1812 and attended lectures by renowned English chemist and physicist Humphry Davy (1778-1829) and John Tatum (1772-1858), founder of the City Philosophical Society. Faraday produced a 300-page book of notes from the lectures and sent it to Davy, who was impressed enough to employ him as a secretary and later as a chemical assistant at the Royal Institution.

Faraday, who was not considered a gentleman in the British class-based society, also served as Davy's valet on a long tour between 1813 and 1815. Made to travel outside the coach and eat with the servants, Faraday was miserable, but the trip allowed him access to the European scientific elite and he eventually was able to conduct his own research. In 1824, he became a member of the Royal Society and the following year was appointed director of the laboratory. In 1833, he was appointed to the Fullerian Professorship of Chemistry, a position he held until the end of his life.

Although Faraday made significant contributions in chemistry, including the discovery of benzene and the invention of an early version of the Bunsen burner, his most important work was in magnetism and electricity. In 1831, a series of experiments led to the discovery of electromagnetic induction. After demonstrating that a changing magnetic field produces an electric field, he used the principles to construct a device called the electric dynamo, the precursor of the modern power generator. In 1845, he discovered the phenomenon now called the Faraday effect, concerning the relationship of light and electromagnetism.

Faraday's mathematical abilities were elementary, but Scottish mathematician James Clerk Maxwell (1831-1879) later used Faraday's work to develop the equations that underlie all modern electromagnetic phenomena theories. In his scientific papers, Maxwell wrote, "The way in which Faraday made use of his idea… shews him to have been in reality a mathematician of a very high order—one from whom the mathematicians of the future may derive valuable and fertile methods."[1]

In 1832, he was granted an honorary doctor of civil law degree from the University of Oxford. Though he was highly respected, Faraday was a very humble man, rejecting a knighthood and twice refusing the presidency of the Royal Society. He was an active elder in his church throughout his life, and although Faraday considered religion and science to be "two distinct things,"[2] he did not see them as conflicting with one another.

Yet even in earthly matters I believe that "the invisible things of Him from the creation of the world are clearly seen, being understood by the things that are made, even His eternal power and Godhead," and I have never seen anything incompatible between those things of man which can be known by the spirit of man which is within him, and those higher things concerning his future, which he cannot know by that spirit.[3]

Faraday refused an offer to be buried in Westminster Abbey but has a memorial plaque near Isaac Newton's tomb. When he died in 1867, he was interred in Highgate Cemetery and was later joined by his wife, Sarah Barnard, when she passed away in 1879.

Faraday's life and passion for science can best be summed up in his own words: "…I cannot doubt that a glorious discovery in natural knowledge, and the wisdom and power of God in the creation, is awaiting our age, and that we may not only hope to see it, but even be honoured to help in obtaining the victory over present ignorance and future knowledge."[4]

References
1. Maxwell, J. 2003. *The Scientific Papers of James Clerk Maxwell: Volume II*. Mineola, NY: Dover Publications, 358-360.
2. Jones, B. 1870. *The Life and Letters of Faraday: Volume II*. London: Longmans, Green and Co., 195-196.
3. Ibid, 325-326.
4. Ibid, 385.

Man of Science, Man of God:
James Clerk Maxwell

From an early age, James Clerk Maxwell had an astonishing memory and an unquenchable curiosity about how things worked. His first teacher, his mother, encouraged him to "look up through Nature to Nature's God":

> His knowledge of Scripture, from his earliest boyhood, was extraordinarily extensive and minute…. These things were not known merely by rote. They occupied his imagination, and sank deeper than anybody knew.[1]

After growing up mostly on an isolated country estate, young Maxwell entered the Edinburgh Academy in 1841. The other boys made fun of his mannerisms, accent, and wardrobe, but he soon befriended Lewis Campbell (his future biographer) and Peter Guthrie Tait. Both would become notable scholars, and remained his lifelong friends. While at Edinburgh, Maxwell won medals for mathematics and Scripture biography.

At age 14, he wrote *Oval Curves*, a paper on the properties of ellipses and curves. It was presented to the Royal Society of Edinburgh by James Forbes, a University of Edinburgh professor of natural philosophy, since Maxwell was "too young" to present it himself. Maxwell entered the university at age 16 and produced *Rolling Curves*. Once again he was considered too young to present it to the Society, so the paper was read by his mathematics professor, Philip Kelland.

In October 1850, Maxwell left Scotland for Cambridge University, where he accomplished a significant portion of his translation of electromagnetism equations, the work for which he is best known. He also laid out the principles of color combination in *Experiments on Colour*—on which occasion he was finally allowed, in March 1855, to present his own paper to the Royal Society of Edinburgh. He became a fellow of Trinity College that October, and the following year applied for and eventually accepted the Chair of Natural Philosophy at Marischal College in Aberdeen.

When the college merged with the University of Aberdeen's King's College in 1860, there was no need for two chairs of natural philosophy, so Maxwell was laid off. He lost an Edinburgh professorship to his childhood friend Tait, but was granted the Chair of Natural Philosophy at King's College in London.

His color research garnered Maxwell election into the Royal Society of London in 1861.

Who: James Clerk Maxwell
What: Father of Electromagnetic Theory
When: June 13, 1831 – November 5, 1879
Where: Edinburgh, Scotland

He often lectured at the Royal Institution, where he regularly conversed with Michael Faraday. At King's College, he produced his most significant work in electromagnetism, a multi-part paper called *On Physical Lines of Force*. He also published papers on electrostatics and displacement current, the latter focusing on the phenomenon known as the Faraday effect.

He resigned from King's College in 1865 and returned to his childhood home at Glenlair, where he wrote the textbook *Theory of Heat* and an elementary treatise called *Matter and Motion*. In 1871, he became the first Cavendish Professor of Physics at Cambridge. He died at 48 in Cambridge of abdominal cancer on November 5, 1879.

Darwin's *Origin of Species* was published during Maxwell's lifetime. Maxwell was not convinced evolution was a viable theory of origins, nor was he afraid to speak on the matter:

> No theory of evolution can be formed to account for the similarity of molecules, for evolution necessarily implies continuous change, and the molecule is incapable of growth or decay, or generation or destruction.…Science is incompetent to reason upon the creation of matter itself out of nothing.[2]

Maxwell is to this day held in high regard in the scientific community, but few know or acknowledge his strong Christian roots or his faith in the authority of God's Word. Virtually every part of his brief, but remarkable, life was spent exploring the wonder of God's creation.

References
1. Campbell, L. and W. Garnett. 1882. *The Life of James Clerk Maxwell: With Selections from His Correspondence and Occasional Writings.* London: Macmillan and Co., 32.
2. Ibid, 359.

Man of Science, Man of God:
Gregor Johann Mendel

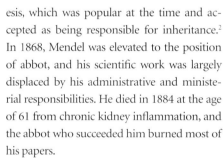

Who: Gregor Johann Mendel
What: Father of Modern Genetics
When: July 20, 1822 – January 6, 1884
Where: Heinzendorf, Hapsburg Empire (Modern-Day Czech Republic)

Gregor Mendel was an Austrian-born, German-speaking Augustinian monk who is famously known as the founder of the modern study of genetics, though his work did not receive much recognition until after his death.

He was born Johann Mendel to a peasant family in the village of Heinzendorf of the Hapsburg Empire, now known as Hyncice of the Czech Republic. As a child, he worked alongside his father to improve the family orchards by grafting, a practice encouraged and sponsored by the land's feudal proprietor, the Countess Maria Truchsess-Ziel.[1] Since grafting was a particular "art" that produced both desirable and undesirable results, working in the orchards introduced the young Mendel to the beginnings of his experimental botany work.

When he was 11, Mendel's schoolmasters recognized his talent for learning and convinced his parents to let him pursue a higher education and, hopefully, a better vocation than the harsh life of a farmer. His parents were frequently unable to cover all of his living expenses in addition to paying his tuition, and he had to work to feed himself. Often he had to do without. In 1838, his father suffered a severe injury that prevented him from doing hard physical labor. The younger Mendel, at 16, then had to entirely support himself and his education.

He graduated from the "gymnasium" at Troppau on August 7, 1840, with high honors, taking first place in all his examinations. Between 1840 and 1843, he studied at the Philosophical Institute in the nearby city of Olomouc. Then, upon recommendation of his physics teacher, Father Franz, he entered the Augustinian Abbey of St. Thomas in Brünn (present-day Brno), adopting the name Gregor when he entered the monastic life.

He was sent to the University of Vienna between 1851 and 1853 to study botany, zoology, chemistry, and physics, and returned to the abbey in Brünn to teach. Between 1856 and 1863, he cultivated some 29,000 pea plants (*Pisum sativum*). The study showed that out of four plants, one received recessive alleles, two were hybrids, and one had the dominant alleles. His experiments were the foundation for two generalizations known today as Mendel's Laws of Inheritance. Based on his work, he produced the paper *Experiments on Plant Hybridization* and read it to the Natural History Society of Brünn in 1865. The society published the paper in its *Proceedings* in 1866.

Mendel's paper was rejected at first, since he evidently produced it as a counter to Darwin's theory of pangenesis, which was popular at the time and accepted as being responsible for inheritance.[2] In 1868, Mendel was elevated to the position of abbot, and his scientific work was largely displaced by his administrative and ministerial responsibilities. He died in 1884 at the age of 61 from chronic kidney inflammation, and the abbot who succeeded him burned most of his papers.

In 1900, Mendel's work was rediscovered and is now the foundation of the science of genetics. In the past hundred years or so, his work has still received criticism and some have gone so far as to accuse Mendel of scientific fraud, even though his experiments have been recreated with the same results. Others have tried to shoehorn his work into the theory of modern evolutionary synthesis, which combines Mendelian genetics with natural selection and gradual evolution.

Though Mendel cannot speak for himself today and defend his work, his life as a priest testifies to his faith in the Creator God. After all, it is one thing to make confession in published books and papers, and it is another to dedicate one's life to those convictions.

References
1. Mendel, G., A. Corcos, and F. Monaghan. 1993. *Gregor Mendel's Experiments on Plant Hybrids: A Guided Study*. New Jersey: Rutgers University Press.
2. See Bishop, B. E. 1996. Mendel's Opposition to Evolution and to Darwin. *The Journal of Heredity*. 87 (3): 205-213.

Man of Science, Man of God: Louis Pasteur

Who: Louis Jean Pasteur
What: Father of Modern Microbiology
When: December 27, 1822 – September 28, 1895
Where: Arbois, France

Little was known about preventative medicine and the causes of disease in the days of Louis Pasteur. Today, we owe all the discoveries in the fields of microbiology and immunology to his work.

Pasteur came from a long line of peasants. Biographers Albert Keim and Louis Lumet wrote that Pasteur was a "rather slow" student and "gave no indication of brilliant qualities. He studied diligently, but without enthusiasm."[1] His father, however, wanted him to become a professor, so the young Pasteur applied himself to his studies and eventually gained admittance to the prestigious École Normale Supérieure, to which he returned in 1856 as director of scientific studies.

Pasteur was a devout Catholic and was "regarded as conforming with the biblical account of the creation."[2] At the time, the concept of spontaneous generation was widely accepted, which maintained that life was generated by non-life (i.e., maggots appeared to arise out of exposed animal carcasses). Darwin used this theory, also known as abiogenesis, to propose that the first life forms miraculously grew out of a "warm little pond, with all sorts of ammonia and phosphoric salts, lights, heat, electricity, etc."[3]

Pasteur conducted experiments comparing organic material that was exposed to air with organic material that was not. Nothing grew in the sealed or filtered vessels. This empirically demonstrated that the fermentation in the open containers was caused by the growth of microorganisms from the air, not spontaneous generation, thus proving biogenesis (life begets life). His studies led to the practices of sterilization and "pasteur"ization, particularly in the fields of medicine and food preparation. He wasn't the first to propose the theory that germs cause disease, but his experiments supported it and he shares the title of "the father of germ theory and bacteriology" with German physician Robert Koch.

Pasteur received criticism for his work and detail-oriented methods. He wouldn't make any claim until he had re-tested several times and was sure of the results. His vigorous approach to his work eventually led to perhaps his most significant contributions: the anthrax and rabies vaccines. Anthrax was responsible for destroying whole flocks of sheep in France, and Pasteur successfully demonstrated the survival of inoculated sheep from the vaccines he developed.

In 1880, Pasteur and his pupils began research into hydrophobia, or rabies. At the time, treatments included cauterization with hot irons, stifling sufferers between two mattresses, and other useless "remedies." Pasteur isolated the rabies virus and developed both a non-virulent and a virulent vaccine. The non-virulent version was given to a nine-year-old boy bitten by a mad dog in 1885. The vaccine succeeded, as it did in subsequent cases. On October 26, 1885, Pasteur officially presented his research on hydrophobia, preventative vaccines, and vaccination treatment applied after infection, and this work was highly praised.

On June 4, 1887, the Pasteur Institute was established in Paris so Pasteur and his researchers could expand their work on infectious diseases.[4] A university was also named for him, and he received high honors and acclaim from his peers, years after he had endured their criticism. Later in life, he suffered from a series of strokes and died from complications in 1895 near Paris.

Pasteur's work set the foundation for some of the most important advances in our modern world. He was an experimentalist of the highest order, and his science was undoubtedly fueled by his faith:

> Are science and the passionate desire to understand anything else than the effect of that spur towards knowledge which the mystery of the universe has placed in our souls? Where are the true sources of human dignity, of liberty, of modern democracy, unless they are contained in the idea of the infinite, before which all men are equal?[5,6]

References
1. Keim, A. and L. Lumet. Translated by Cooper, F. T. 1914. *Louis Pasteur*. New York, NY: Frederick A. Stokes Company, 5-6.
2. Ibid, 63.
3. Darwin, C. Written in 1871, published in 1887. *The Life and Letters of Charles Darwin, including an autobiographical chapter*, vol. 3. London: John Murray, 18.
4. *The history of the Pasteur Institute*. Posted online at www.pasteur.fr/english.html, accessed September 22, 2008.
5. Keim and Lumet, *Louis Pasteur*, 143.
6. Galatians 3:28.

Man of Science, Man of God: George Washington Carver

Who: George Washington Carver
What: Father of Modern Agriculture
When: 1864 or 1865 – January 5, 1943
Where: Diamond Grove, Missouri

Probably no other scientist has had to face as many social barriers as George Washington Carver, the black American botanist noted for revolutionizing agriculture in the southern United States. He was born towards the end of the Civil War to a slave family on the farm of Moses Carver. As an infant, he and his mother and sister were kidnapped by Kentucky night raiders.

It's unclear what happened to his mother and sister, but George was rescued and returned to the Carvers, who raised him and his brother James. He grew up in a deeply segregated world, and very few black schools were available in the South. But his desire for learning prompted him to persevere, and he earned his diploma from Minneapolis High School in Minneapolis, Kansas.

Entering college was even more difficult, but he was eventually accepted at Simpson College in Indianola, Iowa, to study art. In 1891, he transferred to Iowa State Agriculture College in Ames (now Iowa State University) to study botany, where he was the first black student and later the first black faculty member. While there, he adopted the middle name "Washington" to distinguish himself from another George Carver. He received his undergraduate degree in 1894 and his masters in 1896, and became a nationally recognized botanist for his work in plant pathology and mycology. After receiving his masters, he joined Booker T. Washington at the Tuskegee Normal and Industrial Institute (later Tuskegee University) in Alabama to teach former slaves how to farm for self-sufficiency.

Carver revolutionized agricultural science with his cultivation of soil-enriching crops, such as peanuts and soybeans, to revive earth that had been depleted of nutrients from cotton farming. He discovered over 100 uses for the sweet potato and 300 uses for the peanut, including beverages, cosmetics, dyes and paints, medicines, and food products. He conducted numerous research projects that also contributed to medicine and other fields, and used his influence to champion the relief of racial tensions.

He was offered many honors and substantial wealth from patents, but Carver chose not to patent his discoveries: "One reason I never patent my products is that if I did it would take so much time, I would get nothing else done. But mainly I don't want my discoveries to benefit specific favored persons."[1]

Frugal in finance and humble in character, Carver was undoubtedly a deeply devoted Christian. He attributed inspiration of his work to God,[2] and his studies of nature convinced him of the existence and benevolence of the Creator: "Never since have I been without this consciousness of the Creator speaking to me….The out of doors has been to me more and more a great cathedral in which God could be continuously spoken to and heard from."[3]

Carver died January 5, 1943 of complications from injuries he incurred in a bad fall. His life savings of $60,000 was donated to the museum and foundation bearing his name. The epitaph on his grave on the Tuskegee University campus summarizes the life and character of this former slave, man of science, and man of God: "He could have added fortune to fame, but caring for neither, he found happiness and honor in being helpful to the world."

References
1. Carver Quotes. Posted on the George Washington Carver National Monument website at www.nps.gov/gwca.
2. Carver is quoted as saying, "I never have to grope for methods. The method is revealed at the moment I am inspired to create something new. Without God to draw aside the curtain I would be helpless." Federer, W. J. 1994. *America's God and Country Encyclopedia of Quotations.* Coppell, TX: FAME Publishing, 96.
3. Ibid, 97.

A "Short" List of Peanut By-Products Discovered by G. W. Carver:

- Peanut Punch
- Peanut Beverage Flakes
- All Purpose Cream (cosmetic)
- Antiseptic Soap
- Baby Massage Cream
- Face Bleach and Tan Remover
- Facial Lotion
- Facial Powder
- Glycerine
- Hand Lotion
- Peanut Oil Shampoo
- Shaving Cream
- Tetter and Dandruff Cure
- Vanishing Cream
- 30 different Dyes for Cloth
- 19 different Dyes for Leather
- 17 different Wood Stains
- Hen Food (from the peanut hearts)
- 3 different kinds of Stock Food
- Bar Candy
- Caramel
- Chili Sauce
- Chocolate Coated Peanuts
- Curds
- Dry Coffee
- Flavoring Paste
- Meat Substitutes
- Peanut Brittle
- Peanut Cake
- Peanut Flour
- Peanut Popcorn Bars
- Peanut Relish
- Peanut Tofu Sauce
- Salad Oil
- Vinegar
- Worcestershire Sauce
- Castor Oil Substitute
- Emulsion for Bronchitis
- Iron Tonic
- Laxatives
- Axle Grease
- Charcoal (from the shells)
- Diesel Fuel
- Gasoline
- Glue
- Insecticide
- Linoleum
- Lubricating Oil
- Nitroglycerine
- White Paper (from the vines)
- Printer's Ink
- Plastics
- Rubber
- Laundry Soap
- Sweeping Compound

Man of Science, Man of God:
HENRY M. MORRIS

Who: Henry Madison Morris
What: Father of Modern Creation Science Movement
When: October 6, 1918 – February 25, 2006
Where: Dallas, Texas

Henry M. Morris is widely recognized as the founder of the modern creation science movement. He lectured and wrote extensively in defense of a literal interpretation of the Bible's first book, Genesis—particularly the first 11 chapters that describe the creation of the world and all living things, the great Flood of Noah's age, and the human dispersion at Babel.

Background

Dr. Morris was born in Dallas, Texas. He graduated from Rice University in Houston in 1939 with a bachelor's degree in civil engineering and married Mary Louise in 1940. He worked as a hydraulic engineer until 1942, when he returned to Rice to teach civil engineering for the next four years. After this, he worked at the University of Minnesota, where he received his master's degree in hydraulics in 1948 and his Ph.D. in hydraulic engineering in 1950.

In 1951, he became a professor and chair of civil engineering at the University of Louisiana at Lafayette. He then served as a professor of applied science at Southern Illinois University and then as the department chair of civil engineering at the Virginia Polytechnic Institute and State University (Virginia Tech).

Shortly after Dr. Morris received his bachelor's degree from Rice, he accepted the Bible—from Genesis to Revelation—as the infallible and inspired Word of God. In 1946, he published a short book, *That You Might Believe*, exposing the scientific weaknesses in evolution.

The Genesis Flood

In 1961, Dr. Morris and Old Testament expert Dr. John C. Whitcomb published *The Genesis Flood*, the book that was widely acknowledged even by prominent evolutionary paleontologist Stephen J. Gould as "the founding document of the creationist movement."[1]

In it, they unabashedly affirmed their faith in the inerrancy and infallibility of the verbally inspired Word of God and showed the inadequacies of uniformitarianism and evolutionary theory. Drawing on data from the disciplines of hydrology, geology, and archaeology, Drs. Morris and Whitcomb demonstrated how science affirms the biblical record of the great Deluge during the days of Noah.

While Charles Darwin's 1859 *On the Origin of Species* had attempted to provide an explanation—albeit based on imagination instead of science—for the origin of some animals by natural processes instead of by God, *The Genesis Flood* gave a bold, fresh perspective on how the scientific study of natural phenomena in our world is actually consistent with what we read in Scripture.

All of these Biblical references from the Flood record are clearly supported in at least a general way by the actual records of the rocks. Almost all of the sedimentary rocks of the earth, which are the ones containing fossils and from which the supposed geologic history of the earth has been largely deduced, have been laid down by moving waters....Sedimentary rocks by definition are those that have been deposited as sediments, which the Oxford Universal Dictionary defines as "earthy or detrital matter deposited by aqueous agency." Obviously these great masses of sediments must first have been eroded from some

> "[Henry Morris was] the most important creationist of the 20th century, much more so than William Jennings Bryan."[2]
> — Eugenie C. Scott,
> Executive Director of the National Center for Science Education

previous location, transported, and then deposited (perhaps, of course, more than once)—exactly the sort of thing which occurs in any flood and which we have seen must have occurred on a uniquely grand scale during the great Flood of Genesis.[3]

[T]he evidence of the reality of these great events, the Creation and the Deluge, is so powerful and clear that it is only "willing ignorance" which is blind to it, according to Scripture![4]

Dr. Whitcomb, who read Dr. Morris' *That You Might Believe* in 1948 while studying paleontology at Princeton University, said that *The Genesis Flood* would not have been nearly as effective had it been written only by a theologian. "It needed a scientist. And that scientist was Henry Morris," he said in a recent lecture.

Dr. Whitcomb described the difficulties in initially finding a publisher for the book due to its size and subject matter. Nevertheless,

Henry M. Morris continued

> "Dr. Morris is one of my heroes of the faith. He is the man of the Lord raised up as the father of the modern creationist movement…. This is certainly the end of an era in the history of Christendom."[6]
> — Ken Ham, President and CEO of Answers in Genesis

he praised God that it came to print because of the lives that it has changed. For instance, Kitty Foth-Regner dedicated her book *Heaven Without Her* "to Dr. John Whitcomb and the late Dr. Henry Morris, for showing me the truth about where we came from, what we're doing here, and where we're going."[5]

The Institute for Creation Research

In 1963, Dr. Morris and nine other young-earth creationists, including Dr. Duane T. Gish, founded the Creation Research Society. He resigned from his post at Virginia Tech in 1969 and in 1970 founded the Institute for Creation Research as the research division of Christian Heritage College (now San Diego Christian College).

ICR's goal was research, communication, and education in those fields of science that are particularly relevant to the study of origins. In 1981, after receiving approval from the state of California to grant masters degrees in science education, ICR became an autonomous entity.

While serving as ICR's president, Dr. Morris collaborated with scientists and theologians around the world. He wrote more than 60 books on topics that include creation science, evolution, and Christian apologetics, and he lectured worldwide at conferences, churches, and universities. He participated in over 100 debates—many alongside biochemist and ICR vice president Dr. Gish—with evolutionary scientists such as biologist Kenneth R. Miller, zoologist Hubert Frings, and paleontologist David B. Kitts.

Nearly 40 years after its inception, ICR continues to conduct research from the scientific and biblical creation perspective and communicate the truth of God's Word that is found in God's creation.

Other Writings

Dr. Morris wrote extensively on creation science and evolution, producing definitive works such as *Scientific Creationism* (1974), *The Genesis Record* (1976), *The Revelation Record* (1983), *The Biblical Basis for Modern Science* (1984), *Science and the Bible* (1986), and *Biblical Creationism* (1993).

He also addressed Christian apologetics in books such as *Many Infallible Proofs* (1974) and *The Long War Against God* (1989), as well as annotations in *The New Defender's Study Bible* (1995).

In his final book, *Some Call It Science* (2006), Dr. Morris revealed the religion behind the so-called science of the evolutionary establishment. He wrote:

> During the past century…the gospel of new life in Christ has been replaced by the Darwinian "gospel of death," the belief that millions of years of struggle and death has changed pond scum into people and that evolutionary progress will continue inexorably toward heaven on earth.[7]

He then asked the question, "Is it science that supports evolution and disproves the Bible or is it 'science falsely so called'?"[8] He proceeded to present the true religion behind Darwinism as professed by the direct words of some of its most ardent followers, including Stephen J. Gould, P. J. Darlington, Richard Dawkins, Isaac Asimov, and even Charles Darwin himself.

> The faith of the evolutionist…is a splendid faith indeed, a faith not dependent on anything so mundane as evidence or logic, but rather a faith strong in its childlike trust, relying wholly on omniscient Chance and omnipotent Matter to produce the complex systems and mighty energies of the universe. The evolutionist's faith is not dependent on evidence, but is pure faith—absolute credulity.[9]

The evolutionary belief system is antithetical to the gospel, and Dr. Morris warned Christians not to accept "another gospel" and compromise it with creation.[10]

> Any other gospel is another gospel and is not the true gospel. Without the creation, the gospel has no foundation; without the promised consummation, it offers no hope; without the cross and the empty tomb, it has no saving power.[11]

Later Years

Dr. Morris officially retired in January 1996 and took the position of President Emeritus, leaving the leadership roles of ICR to his sons Henry M. Morris III, D. Min. and John D. Morris, Ph.D. He continued to write, producing books, *Days of Praise* devotionals, and articles for ICR's monthly magazine, *Acts & Facts*. Even though he was retired "on paper," his daughter and ICR librarian Mary Smith said, "He was in the office every day until the day he went to the hospital."

After suffering a series of strokes, on February 25, 2006—at the age of 87 and after a full life devoted to the defense of the gospel—Dr. Morris left the hospital in Santee, California, and entered into the joy of the Lord.

References

1. Schudel, M. Henry Morris; Intellectual Father of 'Creation Science.' *The Washington Post*. Posted on washingtonpost.com on March 1, 2006, accessed December 12, 2008.
2. Scott, E. March 4, 2006. Quoted in Rudoren, J. Henry M. Morris, 87, a Theorist of Creationism, Dies. *The New York Times*.
3. Morris, H. M. and J. C. Whitcomb. 1961. *The Genesis Flood: The Biblical Record and Its Scientific Implications*. Phillipsburg, NJ: Presbyterian and Reformed Publishing Company, 124.
4. Ibid, 453.
5. Foth-Regner, K. 2008. *Heaven Without Her*. Nashville, TN: Thomas Nelson.
6. Ham, K. February 25, 2006. Dr. Henry Morris has died. Posted to blogs.answersingenesis.org.
7. Morris, H. 2006. *Some Call It Science*, revised ed. Dallas, TX: Institute for Creation Research, 7.
8. Ibid, 7.
9. Ibid, 22.
10. 2 Corinthians 11:4.
11. Morris, *Some Call It Science*, 50.

For Further Study

Those who wish to use the biographical profiles of *Thinking God's Thoughts After Him* in an educational setting will benefit from the following study pages, which can be reproduced for individual students. Teachers can visit icr.org/great-scientists to access answers for the study questions.

ICR also provides teachers in both Christian and public education valuable online resources through our new Evidence for Creation site. Covering subjects of science, truth, nature, the Bible, and God as Creator, this site allows teachers and students the ability to study the issues of creation, the Bible, and science systematically or by topic. Teachers who want to be prepared with answers for students and colleagues will find the Evidence for Creation site a daily resource in lesson preparation. Visit icr.org/Evidence for more details.

For nearly four decades, the Institute for Creation Research has been providing educational materials to benefit churches, teachers, schools, and individuals who wish to have a greater understanding of the sciences from a biblical creationist worldview. ICR equips believers with evidence of the Bible's accuracy and authority through scientific research, educational programs, and media presentations, all conducted within a thoroughly biblical framework. Our free monthly magazine, *Acts & Facts*, offers fascinating articles and current information on creation, evolution, and more. Our quarterly devotional booklet, *Days of Praise*, provides daily biblical "meat" to strengthen and encourage the Christian witness. Visit icr.org to subscribe to ICR's free publications, or to search our vast archive of articles on a wide variety of biblical and scientific topics.

For an answer key to the study questions, visit icr.org/great-scientists

Thinking About GALILEO GALILEI

1) The *Galileo* spacecraft surprised evolutionary astronomers in 1999 when it discovered far more of the inert gases xenon, krypton, and argon in Jupiter's atmosphere than they expected. This gaseous composition contradicted their theory that Jupiter was built by randomly colliding comets and space rocks, which have much smaller amounts of these three gases. Explain how the fact that Jupiter has an almost circular (only slightly elliptical) orbit around the sun is also inconsistent with the comet collision concept.

Galileo lived from _____ to _____

2) In Galileo's quote on page 5, he refers to two ways of knowing about God. One is general revelation (what can be known through what God has made), and the other is special revelation (what can be known only through the Bible). In the spaces below, write the portion of Galileo's quote that refers to each type of revelation, and then think of an example.

	General Revelation	**Special Revelation**
Galileo's quote:		
Example:		

3) Galileo studied comets. These balls of ice defy an evolutionary explanation because they travel in highly elliptical orbits and have no obvious natural source of origin. The Oort cloud was proposed as a source several decades ago, but there remains no evidence for it. Comets point to a young universe because as they travel, especially near the sun, they lose mass. At the currently-observed rate of erosion, comets should only last about 10,000 years—not even close to the five billion years that has been proposed for their age (see "Evidence for a Young World" by Dr. Russell Humphries, *Acts & Facts,* June 2005, available online at icr.org.). Use an appropriate resource to discover how many earth years pass between each full orbit of Halley's comet:_____ years.

Referring to Genesis 1:14, suggest possibilities regarding God's purpose for comets.

4) Galileo invented a thermometer. Though primitive by today's standards, he recognized the importance of having a standardized, consistent means to measure temperature. Suggest two advantages of temperature standardization and measurement.

a) _____

b) _____

5) Ideally, scientists should be free to follow the evidence where it leads without fear of persecution. Galileo was persecuted by his fellow scientists, just as many modern scientists have lost their jobs merely because they questioned the reigning origins concept: evolution. Read 1 Corinthians 1:18 and 2 Peter 3:1-9, then comment on why you think this persecution should be expected by Christians.

Thinking About Johann Kepler

Kepler lived from _____ to _____

1) Kepler was involved in a number of scientific revolutions. One was the shift of astronomy from a mathematical pursuit to a branch of physics. Another was the change in thinking from geocentrism to heliocentrism. Describe each of these models in your own words.

Geocentrism: _____

Heliocentrism: _____

2) Prior astronomers thought of planets as travelling along circular paths, but this did not match observations. Kepler proposed that planets travel along ellipses, or special ovals. To construct an ellipse, push two thumbtacks 7cm apart through a sheet of paper into a piece of corrugated cardboard. Lay a piece of string on the paper, tucking it under the tacks in a loose loop. Keeping tension on the string, use it as a template to trace an oval onto the paper with a pencil. On another piece of paper, separate the tacks by 10cm and repeat the procedure.

3) Kepler shared Sir Francis Bacon's view that man can learn about God by studying the natural universe, which he considered a supplemental "book" of the Bible. After reading Psalm 138:2, can you describe a danger that might come with setting nature up as another Bible "book"?

_____
_____
_____

4) If a scientist says something about nature that seems to contradict God's Word, then either the scientist or God is mistaken.

a) List possible causes for this kind of apparent contradiction. _____
_____
_____

b) Describe a possible method to resolve this apparent conflict. Then discuss or compare your method with someone else's.

_____
_____
_____

Thinking About ROBERT BOYLE

Boyle lived from _____ to _____

1) Why did Boyle learn to read the Hebrew language?

   _____

   How would you describe a man who learned Hebrew for that reason?

   _____
   _____

2) Boyle's Law is still studied in every basic chemistry course today. Boyle developed it by studying gases in an airtight cylinder. Look up the meaning of the term "inverse." According to Boyle's Law, at constant temperature, if the pressure upon a gas trapped in an airtight cylinder is doubled, what will happen to the volume of gas inside the cylinder?

   _____
   _____

   If the pressure of 69mL of gas in a cylinder is increased from 1.45atm to 3.22atm, what will the final volume be? (Your instructor may choose to explain how to use the formula $p_1V_1 = p_2V_2$.)

   Answer: _____ ml

3) Consider the title of Boyle's book *Things Above Reason*. Unreasonable ideas involve such things as contradictions and altering standard definitions. Do you think that the Word of God could be unreasonable in these ways?

   _____
   _____

   What Scriptures might speak to this subject?

   _____

   How would the presence of contradictions and ambiguous meanings affect the value or impact of the Bible?

   _____

   Whereas the teachings in Scripture might be perfectly reasonable, Boyle seemed sure that man could not discover God using reason alone. According to Hebrews 11:6, what other faculty must man exercise to come to God?

   _____

   What could a Christian do if he or she encounters what appears to be a contradiction in Scripture? Compare or discuss your answer with another's.

   _____
   _____
   _____

4) One "contradiction" that is commonly cited is the different genealogies that are given for Jesus in Matthew and Luke. This is easily resolved by seeing that one genealogy is that of Mary and the other is for Joseph. Joseph's does not include Mary because it is his lineage, not hers. Mary's does not have her name because it follows a traditional format of giving only the fathers' names.

   Starting with Matthew 1:1 and Luke 3:23-38, write the names of the fathers of Mary and Joseph below.

   Matthew 1:_____   Luke 3:_____

Copyright © 2009 by the Institute for Creation Research

Thinking About ISAAC NEWTON

Newton lived from _____ to _____

1) List eight different disciplines that Newton studied, excelled in, or pioneered.

a) _____ b) _____
c) _____ d) _____
e) _____ f) _____
g) _____ h) _____

2) Newton's 1687 publication of *Principia* described for the first time how things in the universe operate in terms of physical causes and effects. Prior to this, most scientific writings described phenomena in terms of mystical concepts such as "life force," or the four Greek elements: earth, air, fire, and water. Of the inventions that were made possible through "cause and effect" experimental science, list the four that you think most significantly impact your life.

a) _____ b) _____
c) _____ d) _____

3) Light from a single star can be divided into its separate colors by refraction. This refracted light reveals what elements are burning in each star. Thus, the composition of stars can be discovered using three of Newton's specialties: optics, refraction, and astronomy. To date, each of the thousands of different stars analyzed has a unique composition. How does this compare with 1 Corinthians 15:41?

_____
_____

In the currently popular Big Bang model of the universe, each of the estimated 10^{22} stars in our universe developed from one uniform ball of energy. In contrast, the creation model follows Psalm 33:6: "By the word of the Lord were the heavens made; and all the host of them by the breath of his mouth." Which of these models best matches the fact that each star has a unique composition?

_____

Why do you think so?

_____

4) Design and perform a basic experiment to test or demonstrate Newton's Second Law of Motion: "**F** = **ma:** the net **f**orce on an object is equal to the **m**ass of the object multiplied by its **a**cceleration."

5) Newton applied his brilliant mind to the study of God as well as the creation, and called God a "necessary" being. God cannot "not" exist. This attribute of God, His necessity, is both scriptural and logical. How does this attribute differ from every other being?

_____
_____

Another way to express God's "necessity" is to say that His nonexistence is not possible. Read Psalm 90:2, then complete the logic in the blank below.

a) God is changeless, according to the Bible.
b) Ceasing to be is a form of change.
c) Therefore, God cannot cease _____

6) Newton wrote that he was inspired to explain the moon's orbit around the earth because of his contemplation of an apple that fell straight down from a tree. Like the stars, the moon has a unique composition. Its exact mass, velocity of orbit, and distance from the earth are necessary for life to exist on earth. Some evolutionary scientists think that the moon originated from a piece of the earth that broke off, perhaps from a collision with a large asteroid. What would Newton have thought about this theory of the moon's origin? Write at least two sentences that Newton might have used to respond to this theory.

_____
_____

Copyright © 2009 by the Institute for Creation Research

Thinking About CHARLES BELL

1) What is the name for a scientist who studies the structure of the human body?
   _____

2) What do we call the study of the interactions between, and actions of, the structures of the human body?
   _____

Bell lived
from _____
to _____

3) Bell was fascinated by the nervous system, and the Bell-Magendie's Law still holds true after almost two centuries. It describes the arrangement of two types of neurons: sensory and motor. Using a biology text or appropriate reference, draw arrows on the diagram below, indicating the direction of impulse of each neuron type. In the appropriate blanks, describe briefly the function of each type.

Sensory   **Motor**

_____   _____
_____   [spinal cord diagram with   _____
_____   labels: central canal,   _____
_____   grey matter, white matter,   _____
_____   dorsal root of spinal nerve,   _____
_____   ventral root of spinal nerve]   _____

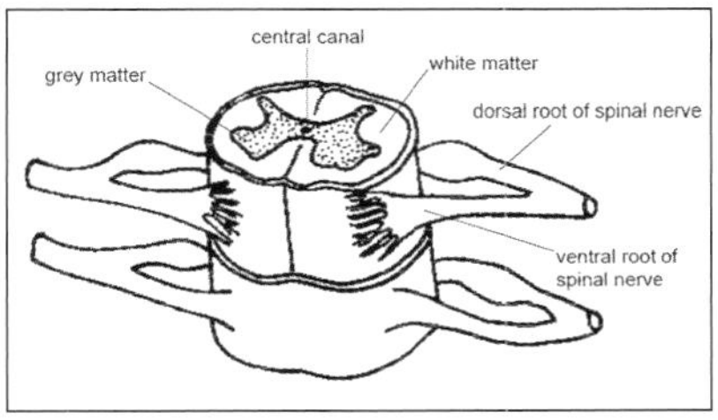

Image credit: Ruth Lawson, Otago Polytechnic

4) We often take our hands for granted, but Bell did not. Human hands were built according to a complex mathematical pattern. Use a ruler to measure your index finger and hand bone lengths in millimeters. First, measure the phalanges. Record the distance between the tip of the first digit to its first knuckle in the blank below. Second, measure the distance from the first to the second knuckle, record it, then again from the second to the third knuckle and record that. Last, measure the distance between the third knuckle and the wrist.

My Data:

First phalange _____ mm

Second phalange _____ mm

Third phalange _____ mm

Metacarpal _____ mm

Averaged Class Data:

First phalange avg. _____ mm

Second phalange avg. _____ mm

Third phalange avg. _____ mm

Metacarpal avg. _____ mm

Collect as many similar measurements from other hands as you can, average them, and record the averages in the appropriate blanks above.

The Fibonacci number series can be constructed by adding together the last two numbers in a list, thus forming the next number of that list. For example, 1+2=3, 2+3=5, 3+5=8, 5+8=13, thus giving the sequence 1, 2, 3, 5, 8, 13, 21, 34, etc.

Adjusting the first phalange average to "1," see how closely the average hand bone measurements fit with a Fibonacci number series.

Copyright © 2009 by the Institute for Creation Research

Thinking About WILLIAM KIRBY

Kirby lived from _____ to _____

1) Kirby observed that insects perform a variety of motions. List six different motions that various insects (which all have six legs) employ in their various modes of locomotion.

   a) _____   b) _____
   c) _____   d) _____
   e) _____   f) _____

2) Scientists recently discovered an instinctive ability of ants. They observed a scout ant encounter a dead grasshopper in a field, return to the mound, and gather a large number of helpers to carry the whole grasshopper back to the mound. When the experiment was repeated with a half and then only a third of a grasshopper body, the number of helpers enlisted to carry the fragments was also a half and a third of that required for the whole grasshopper.

   a) What did the scout ant need to know in order to do this?
   _____
   _____

   b) What basic arguments can be made in *favor* of the ant having taught itself this knowledge?
   _____
   _____

   c) What basic arguments can be made *against* the ant having taught itself this knowledge?
   _____
   _____

3) Kirby reasoned that the instinctive knowledge in insects and other animals came not from themselves, but from their Creator. Today, Kirby's reasoning is automatically labeled unscientific because it refers to God. Do you think this labeling itself is "scientific"? Why or why not?
   _____
   _____

4) Use an appropriate resource to fill in the blanks below.

Insect:	Valuable Function or Product:
(example:) ants	(example:) clean earth's surface
aphids	
mosquitoes	
wasps	
"ladybugs" (lady beetles)	
	produce vanilla
	produce honey and tomatoes

5) Several inventions have been inspired by insect features. See if you can suggest the insect or insect feature that might have influenced the following inventions.

Insect or Feature:	Invention:
	Glow sticks
	Body armor for soldiers
	Camouflage

Copyright © 2009 by the Institute for Creation Research

Thinking About MICHAEL FARADAY

1) Our modern machines are essentially dependent on the electromagnetic principles that Faraday uncovered. Consider the last quote from Faraday given on page 12. If Faraday could visit our world today, in what ways do you think he would feel proud of what later generations did with his knowledge?

_____

_____

Likewise, in what ways do you think he would be disappointed?

_____

_____

Faraday lived from _____ to _____

2) Faraday and Maxwell (see page 13) collaborated in their respective studies on electromagnetism. Work with a friend or partner to construct an electromagnet. You will need one foot of copper wire, a large iron nail or bolt, a D battery, and some paperclips. Strip the insulation off the ends of the wire and tightly and evenly wrap the middle of the wire around the nail about 15 times, as shown in the photo.

Hold one wire tip against one end of the battery, and the other wire tip to the opposite end. Then direct the nail toward the paperclips. See how many paperclips your device can pick up by magnetism. Use a separate piece of paper to record, design and perform an experiment comparing the number of coils around the nail with the number of paperclips the device can lift.

3) Faraday is not generally remembered for his Sunday school lessons, which he taught faithfully for years in his local church in England. What Faraday learned from the Bible affected the way he did science. Examine Proverb 6:6 and list two principles in, or broad assumptions of, this verse that could be applied to scientific investigations.

  a) _____

  b) _____

4) Christianity assumes that the universe is not itself God but is separate from Him, and that it is orderly as a reflection of He who ordered it. Thus, it can be meaningfully investigated. The Christian worldview also holds that time is linear, headed from a beginning toward a conclusion. These are beliefs held by the great scientists described in this book. (See also "Christianity: A Cause of Modern Science?" by Eric V. Snow, *Acts & Facts*, April 1998, available online at icr.org.) What effect might belief in time as a never-ending cycle rather than as a linear progression have on someone's motivation for doing science?

_____

_____

5) Faraday agreed with his church's doctrinal statement: "The Bible, and it alone, with nothing added to it or taken away from it by man, is the sole and sufficient guide for each individual." According to Revelation 22:19, what is the penalty for a person who takes away words from John's book?

_____

_____

Why do you think is it so important to God that His Bible not be added to or taken away from?

_____

_____

Copyright © 2009 by the Institute for Creation Research

Thinking About James Clerk Maxwell

Maxwell lived from _____ to _____

1) Maxwell worked with electricity, which describes the flow of very small, negatively charged electrons. In a car battery, the red post is often labeled "positive," and the black post "negative." Based on this information, describe the general direction of the flow of electrons in a car.

   _____
   _____

2) Demonstrate static electricity by vigorously rubbing an inflated balloon in your hair until it can stick to a wall unaided. Research static electricity, then compose your own description of what causes it.

   _____
   _____

3) Lightning is electricity in the atmosphere, but air is a poor conductor. Job 28:26 says that God "made a decree [law] for the rain, and a way [path] for the lightning of the thunder." It was not until very recently that meteorologists discovered that lightning indeed fills gaps or cracks (paths) in the air, demonstrating the accuracy of the Bible. Research lightning until you can tell another person what causes it. Practice your explanation by writing it below.

   _____
   _____

4) Maxwell's paper *Experiments on Colour* demonstrated that adding certain colors can have effects similar to adding black to shade, or white to tint. Maxwell even assigned numbers to each visually discernable color by using colored paper discs attached to spinning tops. We can quantify and visualize similar color effects by using a standard word processing program. Type some text, highlight it, then click on the color adjustment feature. Find the "custom color" function. Adjust the red and blue hues to 255, and the green to 0. Then raise green's number gradually from 0 to 255. Observe and record what happens to the tint slide bar.

   _____
   _____

5) Maxwell noted that "science itself is incompetent to reason upon the creation of matter itself out of nothing." The idea that science is unable to answer certain real, significant questions is ridiculed by scientism, which is the belief that science is the only way to acquire knowledge. Expressed this way, scientism is self-refuting, and therefore incoherent. How can one have certain knowledge that science is the only way to have knowledge? Proponents of scientism do not "know" their own statement is true by scientific experiments, they just believe it. Beneath each common self-refuting statement below, compose a question that graciously indicates that the statement is incoherent. The first two are included as examples.

   a) "I cannot speak a word of English."
      Q: Is that spoken in English?

   b) "Truth is relative."
      Q: Is that relatively true?

   c) "Meaningful talk of God is impossible."
      Q: Did you mean to talk of God _____?

   d) "Intolerance is unacceptable."
      Q: Is that statement itself very _____?

   e) "There is no reality, only perception."
      Q: Is that reality, or just your _____?

Copyright © 2009 by the Institute for Creation Research

Thinking About Gregor Mendel

1) Mendel's pea experiments were excellent examples of empirical science, partly because he used reductionism, where a trend or pattern is gleaned from a large data set. What is one advantage of using a large data set before drawing conclusions from experiments?

_____
_____

Mendel lived from _____ to _____

2) Mendel knew that some peas were yellow and some green. When he crossed purebred yellow with purebred green, all of the immediate offspring were yellow. When he crossed those yellow offspring with each other, he found that 6,022 were yellow and 2,001 were green. The second generation had 5,474 round and 1,850 wrinkled peas. Estimate the simple, reduced ratio that these larger ratios represent. Begin by dividing the smaller into the larger number in each ratio.

_____

3) Mendel observed that the green trait skipped a generation, being masked by the yellow. In other plants, the wrinkled trait was masked by the smooth. He reasoned that if an individual had two versions of the code for each trait (called alleles) and if one of these was always expressed instead of the other version, then this would explain why certain traits skip a generation and show up in the next. The allele that masks is called dominant, and the allele that gets masked is recessive. An individual with both alleles dominant or both recessive is a homozygote (purebred), and an individual with one of each is a heterozygote.

Choose a letter to represent the pea color trait. Use the capital to represent the dominant allele, and the lowercase for the recessive allele of that trait. Use your letters to calculate the likelihood, expressed as a ratio, that offspring of the following parents will have green peas.

a) A purebred green parent crossed with a heterozygote yellow parent.

_____

b) A heterozygote yellow parent with another heterozygote yellow parent.

_____

c) The cross (b) above represents Mendel's second generation cross, which resulted in the large numbers of offspring reported in study question 2. If the theory of two different alleles is correct, then the predicted ratio should closely match the actual ratio. How do your ratios in questions 2 and 3b compare?

_____
_____

4) In the mid-19th century, the most widely accepted inheritance pattern was called blended inheritance. It held that all traits are the results of blending the parent's traits together. Now we know that there are only a few traits, such as height and skin color in humans, that are largely determined by a blend of the parent genes. However, the pea plant traits that Mendel examined were discontinuous: either tall or dwarf, the flowers either violet or white. Why do you think the rejection of Mendel's results by the scientific community was or was not to be expected?

_____
_____

5) What character qualities do you think Mendel showed by doing his experiments the way he did, and by publishing his conclusions honestly at that time?

_____
_____

Thinking About LOUIS PASTEUR

Pasteur lived from _____ to _____

1) Many believe that in order to truly follow Christ, a person must become a minister or missionary. However, Pasteur and many others today have followed Him by serving as biomedical researchers. Investigate the Christian Medical and Dental Association. Which of the Core Values of the CMDA (available online at cmda.org) did Pasteur's life match? List them here.

_____

_____

2) The biography of Pasteur on page 15 mentions that his work was rejected for many years by the scientific establishment. According to the other biographies in this volume:

a) Were there other great scientists whose achievements were initially rejected? List their names.

_____

_____

b) Many view scientific discovery as a simple additive process, where one generation builds new knowledge upon what came before. How does this view compare with the stories in this book?

_____

_____

3) Researchers have encountered a feature of life that could not have formed by any natural, imagined evolutionary process: the "handedness" of certain chemicals. Living things are made up largely of proteins that are composed of only left-handed amino acids. Outside of living organisms, amino acid "soup" automatically evens out left- to right-handed forms. Trace your left hand onto a piece of paper, then trace your right hand onto a separate page. Collect a class set of these papers and shuffle them. Below, list the information and specifications for the tools that together are required to separate the left-hand papers from the right. (I.e, what exactly is needed to pull out the lefties?)

Information:	Tool specifications:

4) Pasteur made discoveries because of his relentless, thorough, and painstakingly careful procedures. Suppose that Susan tested the effects of lava rock on bean sprout growth by adding lava rock to three sprouting beans. Suppose also that Karen tested the same question by adding lava rock to 30 bean sprouts, but leaving lava rock out of another 30 sprouts. Whose procedures would you trust the results from more, and why?

_____

_____

5) Although Pasteur disproved spontaneous generation, many modern evolutionary scientists still believe that it happened—although they have renamed it "chemical evolution." Atheistic scientists Haldane and Oparin championed this concept during the 1920s, but physicist Freeman Dyson later admitted, "The Oparin picture [of life arising from chemical soup] was generally accepted by biologists for half a century. It was popular not because there was any evidence to support it, but because it seemed to be the only alternative to biblical creationism." (Dyson, F. J. 1985. *Origins of Life*. New York: Cambridge University Press, 31.) Compare John 3:18-21 with Dyson's comment, then suggest a possible reason why scientists from Aristotle through Pasteur's contemporaries, and even today, would reject the principle of biogenesis.

_____

_____

Thinking About GEORGE WASHINGTON CARVER

1) How many uses for the peanut did Carver invent during his life?
   _____

2) What character qualities do you think Carver demonstrated by accomplishing all these inventions from just one plant?

   a)_____ b)_____

   c)_____

3) Have you ever invented new uses for ordinary objects? If so, what?
   _____
   _____

Carver lived from _____ to _____

Challenge: Brainstorm 8 different science demonstrations you could perform using a plastic soft drink bottle. Listed are two ideas to get you started.

   a) make a rocket body   b) build a terrarium

   c)_____   d)_____

   e)_____   f)_____

   g)_____   h)_____

   i)_____   j)_____

4) Peanuts and soybeans are soil-enriching plants called legumes. Use appropriate resources to investigate the role of legumes in the nitrogen cycle and answer these questions:

   a) What chemical is provided by legumes to enrich the soil?_____

   b) Legume plants cannot manufacture this chemical alone. They depend on what specific organism to form a symbiotic infection?_____

   c) What does the legume provide for its symbiont?_____

   d) What does its symbiont provide for the legume?_____

   e) Legumes and their symbionts demonstrate interdependence. Evolution holds that nature designs specific parameters like this all by itself, without God. However, interdependent systems have only been observed to form by intelligent, purposeful agents. What is the most logical answer to the question of where legume symbiosis came from? _____
   _____

5) A web of entities is involved in interdependent ecological relationships with legumes. List four organisms that you think also depend on legumes and the nitrogen cycle.

   a)_____   b)_____

   c)_____   d)_____

6) Because Carver cared as much for the character development as for the intellectual development of his students, he wrote a list of virtues that included the advice "Be clean both inside and out." Describe a procedure, based on Romans 3:20 and 1 John 1:9, to clean our "insides."
   _____
   _____

Copyright © 2009 by the Institute for Creation Research

Thinking About HENRY M. MORRIS

Morris lived from _____ to _____

1) By the time Charles Darwin published his famous book, there was a general consensus that the Bible should not be used or referred to when making scientific observations. Morris broke with that tradition. List two reasons that Henry Morris might have offered to support the idea that the Bible should be referred to when doing science.

   a) _____

   b) _____

2) Morris obtained his doctorate in hydraulic engineering specifically to learn about water movements that occurred during Noah's Flood. Examine Genesis 7 and 8 and answer the following questions.

   a) How long did Noah's Flood last? _____

   b) How long did the rain continue to fall? _____

   c) How extensive were the Flood waters? _____

   d) How long did the waters cover the earth? _____

   e) Were dinosaurs aboard the ark? _____

   f) How many geese were aboard the ark? _____

3) Read 2 Peter 3:5-6. How would a Flood that destroyed the earth have affected pre-Flood places recorded in Genesis 2, such as the land of Havilah or the Garden of Eden?

   _____

   _____

4) Like many of the great scientists featured in this book, Morris enjoyed numbers.

   a) Assuming that an ancient cubit was 18 inches long, calculate the total floor area in all three levels of Noah's ark in square feet. Its dimensions in cubits are L: 300 cubits, W: 50 cubits, H: 30 cubits.

   _____

   b) Assume that the average size of animal cages took up 4 ft^2 of floor space, and that these cages were stacked two high. How many animal cages would two of the three Ark levels have held?

   _____

   c) There were probably fewer than 16,000 different kinds of birds, mammals, and reptiles (including the extinct) that God originally created. Assuming that each of these required its own cage, how many square feet of caging would be required?

   _____

5) Consider the statement "Without the creation, the gospel has no foundation." What does Romans 5:12 give as the cause of death in our world? _____

Henry Morris believed that the majority of sedimentary rocks were formed by the Flood, but those who discount the biblical record insist that these rocks—which contain fossil remains of countless dead creatures—formed over vast eons, long before any Adam, Eve, Garden of Eden, or sin. Consider some consequences to this line of reasoning:

a) The Bible says sin, starting with the first humans, caused death as a just penalty.
b) The Bible says Christ experienced death in our place.
c) Evolutionary scientists say that death (as exhibited by fossils) existed long before humans.
d) Therefore, Christ either:

_____